Yingjie Xing

Electrical properties of PEDOT:PSS film under ultraviolet irradiation

AF153205

Yingjie Xing

Electrical properties of PEDOT:PSS film under ultraviolet irradiation

LAP LAMBERT Academic Publishing

Impressum / Imprint

Bibliografische Information der Deutschen Nationalbibliothek: Die Deutsche Nationalbibliothek verzeichnet diese Publikation in der Deutschen Nationalbibliografie; detaillierte bibliografische Daten sind im Internet über http://dnb.d-nb.de abrufbar.
Alle in diesem Buch genannten Marken und Produktnamen unterliegen warenzeichen-, marken- oder patentrechtlichem Schutz bzw. sind Warenzeichen oder eingetragene Warenzeichen der jeweiligen Inhaber. Die Wiedergabe von Marken, Produktnamen, Gebrauchsnamen, Handelsnamen, Warenbezeichnungen u.s.w. in diesem Werk berechtigt auch ohne besondere Kennzeichnung nicht zu der Annahme, dass solche Namen im Sinne der Warenzeichen- und Markenschutzgesetzgebung als frei zu betrachten wären und daher von jedermann benutzt werden dürften.

Bibliographic information published by the Deutsche Nationalbibliothek: The Deutsche Nationalbibliothek lists this publication in the Deutsche Nationalbibliografie; detailed bibliographic data are available in the Internet at http://dnb.d-nb.de.
Any brand names and product names mentioned in this book are subject to trademark, brand or patent protection and are trademarks or registered trademarks of their respective holders. The use of brand names, product names, common names, trade names, product descriptions etc. even without a particular marking in this work is in no way to be construed to mean that such names may be regarded as unrestricted in respect of trademark and brand protection legislation and could thus be used by anyone.

Coverbild / Cover image: www.ingimage.com

Verlag / Publisher:
LAP LAMBERT Academic Publishing
ist ein Imprint der / is a trademark of
OmniScriptum GmbH & Co. KG
Heinrich-Böcking-Str. 6-8, 66121 Saarbrücken, Deutschland / Germany
Email: info@lap-publishing.com

Herstellung: siehe letzte Seite /
Printed at: see last page
ISBN: 978-3-659-63535-9

Acknowledgments

This book is based on extensive work performed in Peking University. The author wishes to acknowledge many individuals whose help and cooperation aided in the completion of this study. This study is supported by the National Natural Science Foundation of China (Grant nos. 61076057 and 61376059).

Contents

Chapter 1 Introduction to PEDOT:PSS

1.1 Commercial PEDOT:PSS

The rapid development of organic optoelectronics benefits from particular characteristics brought by conducting π-conjugated polymers. These materials show electrical properties of semiconductors with the features of plastics: ease of processing, versatility of chemical synthesis, and flexibility. The success in synthesis of novel conductive polymers offers chances for some fundamental breakthroughs in points of both scientific and industrial view, which was affirmed by the Chemistry Nobel Prize in 2000.

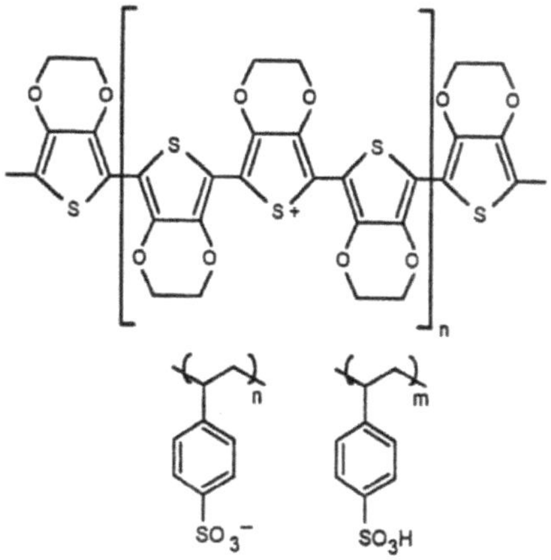

Figure 1.1 Chemical structure of PEDOT (top) and PSS (bottom).

Since the report of doped polyacetylene with a remarkable conductivity in 1977 [1,2], tremendous progresses have been achieved in the synthesis of conductive polymers. In addition to the high conductivity, more and more efforts are made in improving the stability and ease of processing of the polymers, which determines their possible commercial applications to a large extent. Poly(3,4-ethylenedioxythiophene) (PEDOT) is one kind of widely used conductive polymers with satisfied lifetime and production cost. Due to the extremely low solubility of PEDOT in common solvent, the aqueous solution of poly(3,4-ethylenedioxythiophene):poly(styrenesulfonate) (PEDOT:PSS) is developed, which is a big commercial success in the field of utilization of conductive polymers [3]. The chemical structures of PEDOT and PSS are drawn in Figure 1.1. PEDOT:PSS solution opens many opportunities to PEDOT in commercial markets. An industrial scale of PEDOT:PSS is produced every year and used in many practical applications, such as the cathode in solid electrolyte capacitors, antistatic layer in packing films, and transparent electrode or buffer layer in organic optoelectronic devices [3].

Before more technical details of PEDOT:PSS are given in the following sections, the "history" of commercial PEDOT:PSS should be introduced briefly, which is helpful to understand some standards for this kind of conductive polymer. PEDOT was invented by Bayer AG, Leverkusen in 1988 firstly. This material performs high stability, good transparency, and excellent conductivity. PEDOT:PSS was synthesized in 1990 consequently. After the continuous development by Bayer AG and its successors (H. C. Starck GmbH and H. C. Starck Clevios GmbH), a series of PEDOT and PEDOT:PSS with various formulations and conductivities are available for different customers now [3]. Although some other commercial companies also provide this polymer, Clevios GmbH is still among the most popular PEDOT suppliers [4]. Many outstanding characteristics have been achieved by employing commercial Clevios PEDOT products. In order to satisfy the industrial requirement and to compare with the results in literatures, we also use commercial PEDOT:PSS produced by Clevios GmbH (product type: Clevios P VP AI 4083). Most results of PEDOT:PSS mentioned in the following chapters in this book indicate Clevios product, too. Because the aim of this book is not on the synthesis and properties of PEDOT and PEDOT:PSS, only a brief introduction of PEDOT:PSS is depicted in the next part. A detailed description of Clevios PEDOT and PEDOT:PSS can be found in Ref. 3.

PEDOT:PSS is actually a polyelectrolyte complex and the synthesis of PEDOT:PSS is a little different from other conductive polymers. By using PSS as the counterions in water and other oxidizing agents, such as peroxodisulfates and iron(III) salts, PEDOT is polymerized *in situ* under the stabilization effect of PSS [3]. The reaction result is a PEDOT:PSS dispersion, which contains colloidal particles with PEDOT-core surrounding by PSS. In other words, there is no pure PEDOT phase even in the final product. It is reported that various factors, including the ratio of PEDOT to PSS, the size of PEDOT particle, the pH value of the solution, and the concentration of the dispersion, all have influence on the properties of PEDOT:PSS polyelectrolyte complex [3]. Thanks to great efforts made by researchers in Bayer AG and its successors, Clevios PEDOT:PSS has already been optimized and classified to seven types for different purposes. For example, Clevios PEDOT:PSS provides conductivities in a very large range, from 10^{-5} S/cm (product type: P VP CH 8000) to a very high value of 1000 S/cm (product type: PH 1000) [3].

1.2 PEDOT:PSS film

One goal for the well dispersed PEDOT:PSS solution is to provide the possibility of fabricating high quality PEDOT:PSS film in a cost effective manner. To achieve this goal, some aspects need careful consideration and optimization.

The first aspect is the technique to form PEDOT:PSS film. The excellent process-ability of PEDOT:PSS solution makes the formation of PEDOT:PSS film quite easy. Generally, all common "wet" techniques can be employed to deposit PEDOT:PSS film, such as drop casting, spin coating, blade coating. Spin coating is the most often used method in laboratory to prepare small samples, whereas roll-to-roll fabrication system has been realized to print large area PEDOT:PSS layer in a continuous process [5]. Flat PEDOT:PSS films are prepared by these techniques, and high conductivity and good transparency are obtained in these films [3]. In order to improve the coating quality on some substrates (for example, PolyEthylene Terephthalate (PET)), surface modification is conducted to adjust the adhesion property of the substrates before deposition [6]. It is usually proposed that this roll-to-roll fabrication process can reduce the production cost significantly, particularly when comparing with the vacuum deposition technique.

PEDOT:PSS solution is another important aspect that influences the quality, especially the conductivity, of PEDOT:PSS film. As above mentioned, commercial PEDOT:PSS solution with different formulation can be chosen according to the specified usage. Nonetheless, the desire for better property always motivates various attempts to enhance the conductivity of PEDOT:PSS film. A common method is by addition of small amount of polar solvent or high boiling point alcohol into the commercial PEDOT:PSS solution before spin coating, such as dimethylsulfoxide [7], glycerol [8], and sorbitol [9]. Hot PEDOT:PSS solution (70°C) also shows a beneficial effect on the formation of a highly conductive PEDOT:PSS film [10]. Developing new method to improve the conductivity of PEDOT:PSS film always constitutes a large part of PEDOT:PSS research and the progress is still ongoing now. Even though Clevios GmbH supplies commercial products with high conductivity, for example, PEDOT:PSS PH1000 with the conductivity of 1000 S/cm, proper post-treatment can enhance the film conductivity to 3065 S/cm [11]. These values are comparable to that of indium tin oxide (ITO) layer, which is usually prepared by vacuum deposition and used as the transparent electrode now.

Structure of PEDOT:PSS film is a aspect that has been studied for a long time since the invention of PEDOT:PSS. All studies reveal that PEDOT:PSS film is in a amorphous structure and no phase separation on a microscopic scale occurs in PEDOT:PSS film [3]. It is generally believed that this amorphous structure is generated by stacking PEDOT:PSS gels during film formation. Some evidences suggest that particles with the structure of PEDOT core−PSS rich shell distribute evenly in the bulk of the film, whereas a thin PSS rich layer (thickness 3.5 nm) is formed at the film surface [12].

1.3 Interface of PEDOT:PSS film

The interface between PEDOT:PSS film and its adjacent organic layer has a critical influence on hole injection/extraction in organic optoelectronic devices. In this book, we discuss the effect of interface mainly from viewpoint of device physics. Almost all organic optoelectronic devices adopt a planar layered structure, where organic layers are sandwiched between anode and cathode. Generally, one electrode is transparent for light emission/absorption, and ITO is the most common candidate for transparent electrode. As the research on PEDOT:PSS keeps going, good transparency and high conductivity of PEDOT:PSS film attracts more and more interests of using this material in organic optoelectronic devices. Besides some attempts of substituting ITO with high conductivity PEDOT:PSS film, a number of literatures report on using thin PEDOT:PSS film as a anode buffer layer between ITO and organic semiconductor layer. The details of this part will be introduced in the next chapter and only a brief summary is discussed here.

Whether PEDOT:PSS film is used as electrode or anode buffer layer in organic optoelectronic devices, two problems always remain the research focus in this field: how to prepare a interface as expected and how to evaluate the effect of interface on electrical performance of the devices. For the first problem, in order to construct a contact without barrier for charge injection/extraction, most studies use an organic semiconductor with a highest occupied molecular orbital (HOMO) energy closing to the work function of PEDOT:PSS (approximately 5 eV) [13,14]. The answer for the second problem is not as straightforward as it seems to be, though many reports explain the improvement of interracial barrier by analyzing current-voltage curves or measuring photoelectron spectroscopy [15]. There are some contradict conclusions on PEDOT:PSS film with same measuring method [16-18]. This uncertainty brings some doubts about the validity of the solution for the first problem. Therefore, a clear examination of interface of PEDOT:PSS film is needed to clarify above two problems.

As above mentioned, PEDOT:PSS has two main applications in organic optoelectronic devices: electrode and anode buffer layer. The aim of this book is on the latter application: How to improve the conductivity and adjust the work function of PEDOT:PSS film, and how to evaluate the contribution of surface states by analysis of device performance. We hope that our results may help the development of organic optoelectronic devices. We note that the technique used here has not been reported previously, which relies on the ultraviolet (UV) irradiation of PEDOT:PSS film stored in vacuum.

References

[1] C. K. Chiang, C. R. Fincher, Jr., Y. W. Park, A. J. Heeger, H. Shirakawa, E. J. Louis, S. C. Gau, and A. G. MacDiarmid, Electrical Conductivity in Doped Polyacetylene, *Physical Review Letters*, 39(17), 1098–1101, (1977).

[2] H. Shirakawa, E. J. Louis, A. G. MacDiarmid, C. K. Chiang, and A. J. Heeger, Synthesis of Electrically Conducting Organic Polymers: Halogen Derivatives of Polyacetylene, (CH)x, *Journal of the Chemical Society, Chemical Communications*, 578–580 (1977).

[3] Andreas Elschner, Stephan Kirchmeyer, Wilfried Lövenich, Udo Merker, Knud Reuter, PEDOT: principles and applications of an intrinsically conductive polymer, Boca Raton: CRC Press, 2011

[4] Wikipedia, the free encyclopedia, http://en.wikipedia.org/wiki/PEDOT:PSS

[5] Heraeus Clevios™ PEDOT:PSS Technology in printed organic solar, http://www.printedelectronicsworld.com/articles/heraeus-clevios-8482-pedot-pss-technology-in-printed-organic-solar-00005271.asp, 21 Mar 2013

[6] C. Koidisa, S. Logothetidisa, C. Kapnopoulosa, P.G. Karagiannidisa, A. Laskarakisa, N.A. Hastas, Substrate treatment and drying conditions effect on the properties of roll-to-rollgravure printed PEDOT:PSS thin films, Materials Science and Engineering B 176 (2011) 1556– 1561

[7] Na S, Wang G, Kim S, et al. Evolution of nanomorphology and anisotropic conductivity in solvent-modified PEDOT:PSS films for polymeric anodes of polymer solar cells. J Mater Chem, 2009, 19(47): 9045–9053

[8] Snaith H, Kenrick H, Chiesa M, et al. Morphological and electronic consequences of modifications to the polymer anode 'PEDOT:PSS'. Polymer, 2005, 46(8): 2573-2578

[9] Nardes A, Kemerink M, Kok M, et al. Conductivity, work function, and environmental stability of PEDOT:PSS thin films treated with sorbitol. Org Electron, 2008, 9(5): 727–734

[10] Bettina Friedel, Thomas J.K. Brenner, Christopher R. McNeill, Ullrich Steiner, Neil C. Greenham, Influence of solution heating on the properties of PEDOT:PSS colloidal solutions and impact on the device performance of polymer solar cells, Organic Electronics 12 (2011) 1736–1745

[11] Xia Y, Sun K, Ouyang J. Solution-processed metallic conducting polymer films as transparent electrode of optoelectronic devices. Adv Mater, 2012, 24(18): 2436-2440

[12] Hwang J, Amy F and Kahn A 2006 Organic Electronics 7 387

[13] Interface Engineering for Organic Electronics, Hong Ma, Hin-Lap Yip, Fei Huang, and Alex K.-Y. Jen, Adv. Funct. Mater. 2010, 20, 1371–1388

[14] Po R, Carbonera C, Bernardi A, Tinti F and Camaioni N, Polymer- and carbon-based electrodes for polymer solar cells: Toward low-cost, continuous fabrication over large area, 2012 Solar Energy Materials & Solar Cells 100 97

[15] J. C. Bernède, A. Godoy, L. Cattin, F. R. Diaz, M. Morsli and M. A. del Valle (2010). Organic Solar Cells Performances Improvement Induced by Interface Buffer Layers, Solar Energy, Radu D Rugescu (Ed.), InTech

[16] Benor A, Takizawa S, Perez-Bolivar C and Anzenbacher P 2010 Organic Electronics 11 938

[17] Benor A, Takizawa S, Chen P, Perez-Bolivar C and Anzenbacher P 2009 APPLIED PHYSICS LETTERS 94 193301

[18] Helander M G, Wang Z B, Greiner M T, Liu Z W, Lian K and Lu Z H 2009 APPLIED PHYSICS LETTERS 95 173302

Chapter 2 PEDOT:PSS film as a anode buffer layer

2.1 Hole injection layer in organic light emitting diodes

Organic light emitting diodes (OLEDs) have already entered the commercial market for some time. Up to date, sophisticated OLEDs generally adopt a multiple layered structure: an active emitting layer locates at the center and two charge transport layers attach to two faces of the emitting layer for hole-transporting/electron-blocking and electron-transporting/hole-blocking, respectively. The working mechanism of OLEDs can be briefly described as following: When a positive voltage is applied across the OLED, holes and electrons are injected into the device from anode and cathode, respectively; then light emission takes place in the emitting layer via radiative decay of excitons, which are generated by recombination of holes and electrons [1]. In order to obtain a high efficiency from viewpoint of device physics, charge injection from electrodes should be nearly barrier-free and radiative decay of excitons should occur away from the electrodes to reduce exciton quenching by surface plasmons. A buffer layer between electrode and active layer is therefore intentionally inserted to achieve above targets [1]. This layer is usually called hole injection layer (HIL). In practice, one layer can act as the roles of both hole injection layer and hole transport layer sometimes, and a similar case occurs for electron injection layer and electron transport layer. A device structure of OLED is schematically drawn in Figure 2.1.

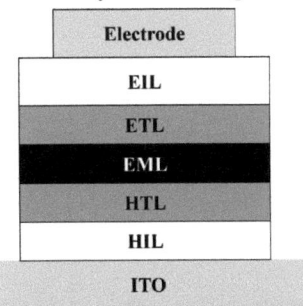

Figure 2.1 Typical device architecture of organic light emitting diodes. ITO stands for indium tin oxide, HIL for hole injection layer, HTL for hole transporting layer, EML for emitting layer, ETL for electron transporting layer, and EIL for electron injection layer.

Various HILs in OLEDs have been studied and reviewed thoroughly in literatures [2]. In practical OLEDs, a direct risk of thin active layer (several tens of nanometer) is electrical short due to the random peaks in ITO substrate. Solution processed PEDOT:PSS layer can eliminate this problem effectively, which is usually considered as one benefit of using PEDOT:PSS film as HIL. Another merit of PEDOT:PSS HIL is the lower barrier for hole injection, which may be the most often mentioned application of PEDOT:PSS in organic electronics [3]. Such an optimization of charge injection can effectively reduce the operating voltage of OLEDs, because the work function of PEDOT:PSS (approximately 5 eV) is larger than that of ITO (approximately 4.5 eV) and is close to the HOMO energies of most wide band-gap hole transporting materials (approximately 5.5 eV). With metal-insulator-metal (MIM) model and the assumption of vacuum level alignment used in numerous literatures, the energy level mismatch and barrier height adjustment are demonstrated in Figure 2.2.

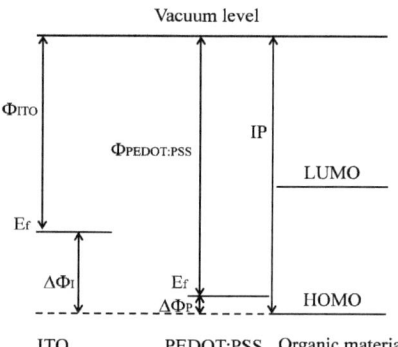

Figure 2.2 Energy diagram of ITO anode with/without PEDOT:PSS hole inject layer and the adjacent organic layer. IP denotes ionisation potential, HOMO denotes highest occupied molecular orbital, and LUMO denotes lowest unoccupied molecular orbital. $\Delta\Phi_I$ and $\Delta\Phi_P$ are the barrier heights for hole injection from ITO or PEDOT:PSS to organic layer.

Among various HILs in OLEDs, Clevios PEDOT:PSS (product type: P VP AI 4083) is one of the most popular candidates to coat ITO anode. A number of literatures employ Clevios P VP AI 4083 as the hole injection layer [4]. We also use this material as the benchmark in our experiments. Clevios P VP AI 4083 shows the lowest viscosity among Clevios PEDOT:PSS products [4]. However, the conductivity of PEDOT:PSS film prepared with this solution is rather low (10^{-3} S/cm). As mentioned in Chapter 1, extensive efforts have been devoted to this problem and some progresses are reported [5-9]. A novel technique to enhance the conductivity of this kind of PEDOT:PSS is developed by UV illumination. Furthermore, the work function of UV treated PEDOT:PSS film is improved by this method. The details of these results will be introduced in the next chapters.

2.2 Anode buffer layer in organic solar cells

Organic solar cells (OSCs) attract worldwide research interests because of the potential for large scale fabrication via cheap wet processes. Organic materials used in OSCs, instead of crystalline silicon in traditional solar cells, offer more possibility to reduce the overall cost, because these function tunable molecules are synthesized by inexpensive chemical routes. However, although considerable progresses have been achieved in the last two decades, the performance of OSCs is still below their inorganic counterparts, particularly in efficiency and lifetime [10]. One reason for this behavior comes from physical properties of organic materials.

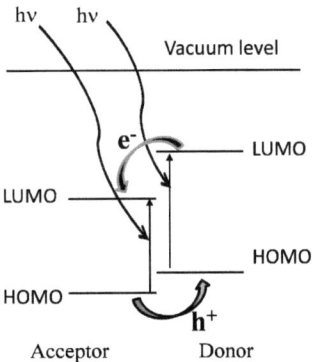

Figure 2.3 Schematic demonstration of exciton dissociation at proper donor/acceptor interface. After excitons are generated in donor and acceptor materials by light irradiation, those excitons near the interface split to unbound holes and electrons by charge transfer across the donor/acceptor interface (electrons from donor to acceptor and holes from acceptor to donor).

Organic semiconductors exhibit a different optical absorption process from inorganic semiconductors. Instead of free electrons and holes generated in inorganic semiconductors by proper irradiation (photon energy larger than the band gap), excitons are produced in organic semiconductors under the same condition [2,3]. The first small molecule OSC in 1986 [11], which utilizes a bilayer structure by stacking electron donor and acceptor materials, reveals the ability of splitting excitons near the donor/acceptor interface to unbounded electrons and holes. A demonstration of exciton dissociation at donor/acceptor interface is schematically drawn in Figure 2.3. Under a suitable field, the separated holes and electrons go through the donor and acceptor layer and then are collected by anode and cathode, respectively. The low efficiency of bilayer small molecule OSCs is therefore easy to understand by small mobility in common organic semiconductors and limited area of planar interface for exciton splitting.

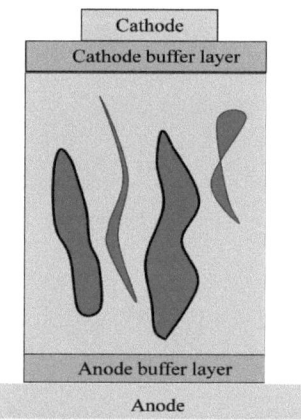

Figure 2.4 Typical device architecture of bulk heterojunction polymer solar cells. A mixture with possible phase segregation is sandwiched between anode buffer layer and cathode buffer layer.

A breakthrough in the field of OSCs is the invention of bulk heterojunction polymer solar cell [12]. This device structure employs a molecule level mixture of p-type polymers and n-type fullerene derivatives to significantly enlarge the interface area, which is usually obtained by spin coating a solution of these two materials. It is obvious that no counterpart concept of bulk heterojunction exists in classic device physics. A device structure of bulk heterojunction polymer solar cells is schematically drawn in Figure 2.4. Since the first report of bulk heterojunction polymer solar cell in 1995 [12], a dramatic improvement of certificated device efficiency (10.7%) has been reported and this kind of device has been considered as one of the most promising candidates of the second generation solar cells, which stands for cheap thin film solar cells with an intermediate efficiency [13].

PEDOT:PSS film is almost a standard anode buffer layer in bulk heterojunction polymer solar cells, no matter which kind of polymer or fullerene is used [4]. Similar as the roles in OLEDs, PEDOT:PSS film is used firstly to avoid the electrical short by random peaks in ITO and to adjust the barrier for hole injection. Alone with research goes through, more beneficial effects are found for PEDOT:PSS as the anode buffer layer in bulk heterojunction polymer solar cell.

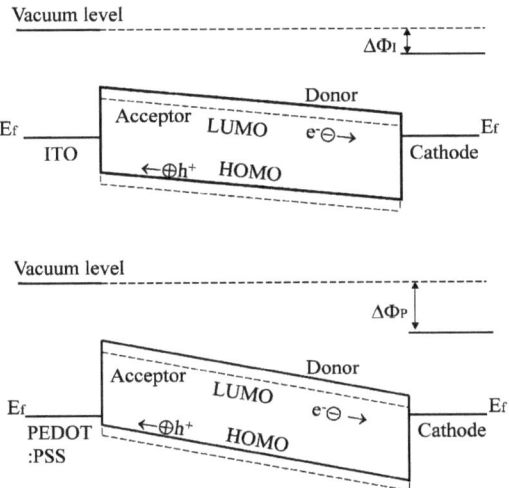

Figure 2.5 Schematic energy diagrams of bulk heterojunction polymer solar cells using ITO without PEDOT:PSS (top) and with PEDOT:PSS (bottom). $\Delta\Phi_I$ (top) denotes the work function difference between ITO and metal cathode. $\Delta\Phi_P$ (bottom) denotes the work function difference between PEDOT:PSS and metal cathode.

Under the illumination of a fixed light intensity (for example, AM 1.5, 100 mW/cm^2), less loss inside the device (including both amount of free charge carriers and potential difference between holes and electrons) means the device outputs more power to the outer circuit. In addition to the injection and extraction processes of charge carriers in ordinary inorganic solar cells, an extra problem needs more careful consideration in bulk heterojunction polymer solar cells: how to efficiently extract holes and electrons to anode and cathode, respectively. Although various methods have been developed to promote an ideal phase separation in the active layer (shown in Figure 2.4), the interpenetrating distribution of polymer and fullerene molecules makes the transport model and extraction routes unclear [14]. A helpful solution for this problem is to supply an intrinsic electric field (direction from anode to cathode) to assist the transport of free charge carriers through the active layer of the solar cell. The work function difference between anode and cathode in OSCs creates such an intrinsic field. In addition to using low work function metals as cathode, such as Ag, Al, or Ca, an anode with high work function is also necessary. Coating PEDOT:PSS (work function approximately 5 eV) on ITO (work function approximately 4.5 eV) improves the work function of anode, resulting in a stronger field to extract charge carriers and a electron blocking layer to avoid a direction connection between anode and cathode by fullerenes. Energy level diagram of bulk heterojunction polymer solar cells is shown in Figure 2.5.

As above discussed, PEDOT:PSS layer as a anode buffer layer is a useful component in organic optoelectronic devices. The need for better devices requires optimization of every part in the devices, including the anode buffer layer. Work function and conductivity are two most important properties of the anode buffer layer. We will introduce our efforts on improving these two properties of PEDOT:PSS film in next chapters.

References
[1] Jan Kalinowski, Organic Light-Emitting Diodes: Principles, Characteristics, and Processes, CRC Press, Boca Raton: 2005.

[2] Hong Ma, Hin-Lap Yip, Fei Huang, and Alex K.-Y. Jen, Interface Engineering for Organic Electronics, Adv. Funct. Mater. 2010, 20, 1371–1388

[3] J. C. Bernède, A. Godoy, L. Cattin, F. R. Diaz, M. Morsli and M. A. del Valle (2010). Organic Solar Cells Performances Improvement Induced by Interface Buffer Layers, Solar Energy, Radu D Rugescu (Ed.), InTech

[4] Andreas Elschner, Stephan Kirchmeyer, Wilfried Lövenich, Udo Merker, Knud Reuter, PEDOT: principles and applications of an intrinsically conductive polymer, Boca Raton: CRC Press, 2011

[5] Na S, Wang G, Kim S, et al. Evolution of nanomorphology and anisotropic conductivity in solvent-modified PEDOT:PSS films for polymeric anodes of polymer solar cells. J Mater Chem, 2009, 19(47): 9045–9053

[6] Snaith H, Kenrick H, Chiesa M, et al. Morphological and electronic consequences of modifications to the polymer anode 'PEDOT:PSS'. Polymer, 2005, 46(8): 2573-2578

[7] Nardes A, Kemerink M, Kok M, et al. Conductivity, work function, and environmental stability of PEDOT:PSS thin films treated with sorbitol. Org Electron, 2008, 9(5): 727–734

[8] Bettina Friedel, Thomas J.K. Brenner, Christopher R. McNeill, Ullrich Steiner, Neil C. Greenham, Influence of solution heating on the properties of PEDOT:PSS colloidal solutions and impact on the device performance of polymer solar cells, Organic Electronics 12 (2011) 1736–1745

[9] Xia Y, Sun K, Ouyang J. Solution-processed metallic conducting polymer films as transparent electrode of optoelectronic devices. Adv Mater, 2012, 24(18): 2436-2440

[10] Pankaj Kumar and Suresh Chand, Recent progress and future aspects of organic solar cells, PROGRESS IN PHOTOVOLTAICS: RESEARCH AND APPLICATIONS, 2012; 20:377–415

[11] Tang CW. Two layer organic photovoltaic cell, Applied Physics Letters 1986; 48: 183–185.

[12] G. Yu, J. Gao, J. C. Hummelen, F. Wudl, A. J. Heeger, Polymer Photovoltaic Cells: Enhanced Efficiencies via a Network of Internal Donor-Acceptor Heterojunctions, Science 1995:Vol. 270 no. 5243 pp. 1789-1791

[13] Martin A. Green, Keith Emery, Yoshihiro Hishikawa, Wilhelm Warta and Ewan D. Dunlop, Solar cell efficiency tables (version 44), Progress in Photovoltaics: Research and Applications, 2014; 22:701–710

[14] Jonathan D. Servaites, Mark A. Ratner, and Tobin J. Marks, Organic solar cells: A new look at traditional models, Energy Environ. Sci., 2011, 4, 4410 –4422

Chapter 3 Tuning work function of PEDOT:PSS film by UV irradiation

3.1 Work function of PEDOT:PSS film

As a common hole injection layer in organic optoelectronics, the work function of PEDOT:PSS film has a significant influence on the injection/extraction barrier and then on final performance of devices. However, it is difficult to adjust the barrier height in practical devices because the work functions of both PEDOT:PSS and common organic materials for active layer are fixed. Although changing material is an option to solve this problem, other optimization of the whole device than the barrier is needed to obtain a satisfied behavior in this case [1]. Tuning work function of PEDOT:PSS film in the required direction is therefore desirable to achieve ideal transport characteristics in optoelectronic devices.

PEDOT:PSS film is usually fabricated on various substrates by spin-coating or printing technique. A PSS-rich layer (thickness of approximately 3.5 nm) is detected at the surface of as-prepared PEDOT:PSS film, caused by vertical phase separation during film formation [2]. Therefore, the work function of PEDOT:PSS film increases slightly because the work function of PSS is higher than that of PEDOT. Further research reveals that a more higher work function may be achieved by enlarging PSS content at the film surface with PSS-enriched PEDOT:PSS solution [3]. A small reduction in the work function (0.1–0.3 eV) may be obtained by removing such a thin PSS-rich layer with a high boiling point solvent or polar solvent [4,5]. Aside from these solution methods, UV treatment is an effective "dry" method of increasing the work function of PEDOT:PSS film. An improvement of 0.2–0.4 eV has been reported in UV ozone–treated PEDOT:PSS [6]. Residual water also influences the work function of annealed PEDOT:PSS film. *In situ* photoelectron spectroscopy has revealed that annealing in high vacuum (1×10^{-10} mbar) enhances the work function of PEDOT:PSS significantly [7]. However, the increased work function reduces when this sample is returned to air, indicating a limitation of this treatment.

Accurate determination of the change in the work function requires a clean sample surface and controllable testing environment. The work function of PEDOT:PSS film has been studied by ultraviolet photoemission spectroscopy (UPS, in high vacuum) [2],

Kelvin probe (KB, in air) [4,5], and photoelectron yield spectroscopy (PYS, in air) [3,6]. Nonetheless, how to detect work function change over time under different atmospheres remains a challenge.

3.2 Field emission measurement of PEDOT:PSS film and Fowler–Nordheim theory

PEDOT:PSS is often treated as a metal in organic electronics. This fact means that some methods for investigating metal can be used to measure work function change in PEDOT:PSS film. Field emission analysis is a highly developed technique used to study the work functions of metals, and Fowler–Nordheim theory has been widely shown to be reliable. We investigate the change in the field emission property and work function of PEDOT:PSS film *in situ*.

Figure 3.1 SEM image of PEDOT:PSS-W tip used in present experiment. A semitransparent PEDOT:PSS layer is coated on the W tip. (Adopted from Ref. 10.)

PEDOT:PSS film is coated on a W tip for field emission measurement. A tungsten tip is fabricated with a thin tungsten wire (diameter 0.1 mm) in NaOH solution with the standard etching method. This tip is mounted in a homemade stainless field emission microscope to test the stability under high voltage. The base pressure of the field emission microscope is approximately 2×10^{-7} Pa. This step is necessary to obtain reliable results because the shape of the W tip may become slightly blunt after continuous emission. Only those tips with stable field emission are chosen as the clean substrate for coating a thin layer of PEDOT:PSS. The deposition of PEDOT:PSS film is

15

realized by dipping the W tip into a droplet of PEDOT:PSS and then drying in air. Lamp heating is used to prepare the tip with solid coverage. This procedure is repeated four times to obtain a PEDOT:PSS layer on the tip with the typical thickness of that in organic optoelectronic devices (several tens of nanometers). Scanning electron microscope (SEM) observation confirms that a thin film with a good shape is formed on the tip surface. We refer to this type of tip as a PEDOT:PSS-W tip. Figure 3.1 shows a SEM image of the PEDOT:PSS-W tip. Complete coating of the sharp W tip with PEDOT:PSS ensures that the field emission property is completely determined by the outer PEDOT:PSS layer. This PEDOT:PSS-W tip is placed in the field emission microscope again for measurement. During the heating process for degassing the system, the heating temperature is always lower than 200°C, to avoid decomposing PEDOT and PSS. The PEDOT:PSS-W tip can be treated in a vacuum chamber by two ways: UV light (wavelength 254 nm, SunMonde, 8 W) irradiation through a quartz window, or oxygen exposure. A small amount of high purity oxygen can be introduced into the chamber via a needle valve, and the pressure of the chamber is monitored with a gauge to control the content of oxygen. The highest pressure during exposure is approximately 10^{-5} Pa. A schematic architecture of the home-made field emission microscope is drawn in Figure 3.2. The consequence of UV irradiation, oxygen exposure and field emission can be chosen as our arrangement with this setup.

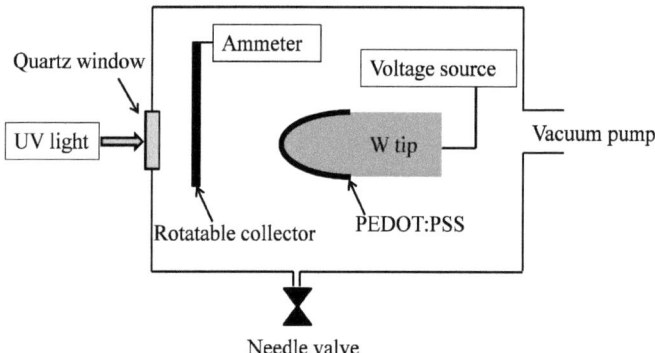

Figure 3.2 Schematic configuration of field emission microscope.

We find that stable field emission is more easily obtained from PEDOT:PSS-W tip than from W tip. Both field emission pattern and field emission current become more stable after performing brief field emission several times. No additional heating is needed to remove the absorption of the PEDOT:PSS-W tip. The position of the PEDOT:PSS-W tip in the chamber remains unchanged throughout each experiment. To protect the PEDOT:PSS-W tip during multiple measurements, the largest emission current is limited to less than 0.1 μA.

With the measurement results, we calculate the change in PEDOT:PSS work function according to Fowler–Nordheim model. Details of Fowler–Nordheim theory can be searched in some classic textbooks [8,9]. In a simplified form of Fowler–Nordheim theory, field emission current is usually expressed as the following equation:

$$J = 1.54 \times 10^{-6} \frac{F^2}{\varphi} \exp\left(-6.83 \times 10^7 \frac{\varphi^{3/2}}{F}\right) \tag{1}$$

where J is the emission current density, F is the field at the tip surface, and φ is the work function of the metal. The field in this equation can be calculate with the formula

$$F = \beta V = \frac{V}{k r_t}$$

where β is the local field conversion factor, r_t is the radius of curvature of the tip, and k is a factor related to the shape of the tip (in most cases, 5 may be used as the value of k) [8]. According to Eq. (1), a linear relationship in the $\ln\left(\frac{I}{V^2}\right) \sim \frac{1}{V}$ plot (F–N plot) can be found when field emission occurs, in which the slope of this straight line is determined by the work function of the metal and the shape of the tip. A better approximation for the slope, that considers the effect of the imaging force, gives

$$s = -6.83 \times 10^7 \frac{\varphi^{3/2}}{\beta} s(y) \tag{2}$$

where $s(y)$ is a slowly varying function with a value of 1 to 0.833; in first-order approximation, 0.917 may be used [8].

Among several early applications of field emission technique, gas absorption on metal surface is studied by investigating field emission performances from clean tip surface and from molecule absorbed surface *in situ*. Significant variation of emission current and pattern are observed during single layer absorption on metal surface [8,9]. The change of surface work function due to gas absorption can be calculated as following: If we assume that the value of β remains constant, variation of the work function ($\Delta\varphi$) is the sole cause of slope change (Δs), according to Eq. (2). This assumption is true in gas absorption experiment because no change in tip shape or anode–cathode distance occurred. The change of work function before and after absorption is calculated using Eq. (3) (φ and s are initial work function and slope, respectively).

$$\varphi + \Delta\varphi = \left(\frac{s+\Delta s}{s}\right)^{2/3} \varphi \tag{3}$$

The results reveal that field emission is a very sensitive technique to detect the change of work function [8,9]. In this chapter, we use this technique to investigate the variation of work function of PEDOT:PSS film under UV illumination *in situ*.

3.3 Effect of UV irradiation and oxygen exposure on work function of PEDOT:PSS film

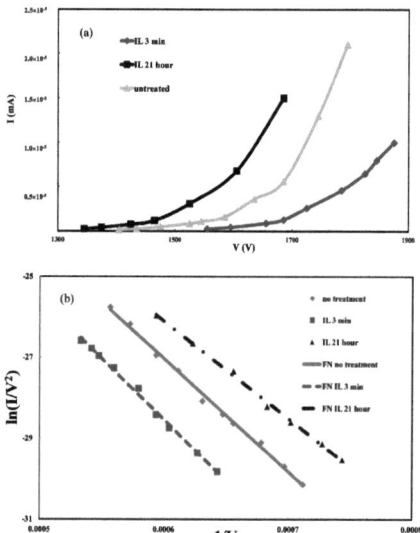

Figure 3.3 (a) Field emission current–voltage (I–V) curves measured after 3 min (diamond) and 21 h (square) of UV illumination. The data of untreated PEDOT:PSS-W tip (triangle) is also shown for reference. (b) Corresponding fitting lines of F–N plot, square for 3 min of UV illumination, triangle for 21 h of UV illumination, diamond for untreated PEDOT:PSS-W tip. IL denotes UV illumination. (Adopted from Ref. 10.)

We measure the original field emission property from untreated PEDOT:PSS-W tip firstly in the dark. Then we irradiate the PEDOT:PSS-W tip using 254 nm UV through a quartz window. After a period of illumination, this window is covered to block the UV light. The field emission is measured again in the dark. This process is repeated approximately ten times to investigate the time dependence of the field emission property of the PEDOT:PSS-W tip on UV irradiation.

TABLE 3.1. Time dependence of slope of F–N line and calculated work function. IL and OX denote UV illumination and oxidation, respectively.

Treatment time	Slope (Absolute value)	Work function (eV)
No treatment	27993	5.06
IL 3 min	30330	5.48
IL 5 min	30445	5.50
IL 8 min	30066	5.43
IL 1 h	28243	5.11
IL 2 h	27488	4.97
IL 3 h	26945	4.87
IL 16 h	25626	4.63
IL 21 h	24019	4.34
OX 1 h	25213	4.56
OX 3 h	27782	5.02

Figure 3.3(a) shows the field emission current–voltage curves after 3 min and 21 h of illumination (the first and last measured point). Figure 3.3(b) shows the fitting lines for the F–N plot, and Table 3.1 lists the slope values. The data on the untreated PEDOT:PSS-W tip is also shown in Figure 3.3 for reference. The slope increases significantly after a short period of UV irradiation (3 min) and decreases afterward (21 h). Using 5.06 eV as the work function of the untreated PEDOT:PSS film, which is measured by PYS separately, the work function of the PEDOT:PSS film after a different period of UV irradiation is calculated using Eq. (3) and is listed in Table 3.1. A high work function (5.50 eV) is measured after 5 min of UV irradiation. Then, a monotonous reduction in the work function is observed through the *in situ* field emission technique. The lowest work function after 21 h of UV illumination is 4.34 eV. This result demonstrates that UV irradiation in vacuum is an effective method of decreasing the work function of PEDOT:PSS film.

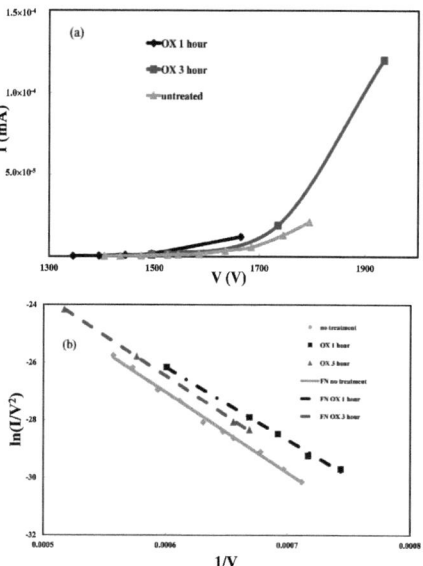

Figure 3.4 (a) Field emission current–voltage (I–V) curves measured after 1 h (diamond) and 3 h (square) of oxygen exposure. The data of untreated PEDOT:PSS-W tip (triangle) is also shown for reference. (b) Corresponding fitting lines of F–N plot, square for 1 h of oxygen exposure, triangle for 3 h of oxygen exposure, diamond for untreated PEDOT:PSS-W tip. OX denotes oxidation. (Adopted from Ref. 10.)

However, PEDOT:PSS is not typically used in a vacuum chamber environment. Although the glove box is used in many laboratories today, a small amount of oxygen leakage or short period of ambient exposure before device encapsulation can be rarely avoided. The stability of a UV-treated PEDOT:PSS surface under ambient atmosphere is a problem that needs attending to. With the abovementioned UV-treated PEDOT:PSS-W tip, we measure the change in the work function caused by oxygen exposure through the *in situ* field emission technique. After 1 h exposure of the UV-treated PEDOT:PSS-W tip, the chamber is again evacuated to 10^{-7} Pa for field emission measurement. This procedure is repeated once more after an additional 2 h of exposure. Particular attention is paid to field emission measurement after oxygen introduction. Only those data points showing stable emission current are recorded and used to calculate the slope. Figure

3.4(a) shows the field emission current–voltage curves after 1 h and 3 h of oxygen exposure and Figure 3.4(b) shows the corresponding F–N fitting lines. The data on the untreated PEDOT:PSS-W tip is also shown in Figure 3.4 for reference. Table 3.1 lists the slope values and calculated work function. Enhancement of the work function due to oxygen exposure is clear, and the work function after 3 h exposure (5.02 eV) recovers almost to its original value (5.06 eV). Figure 3.5 shows the change in the work function under UV irradiation and oxygen exposure. The turning point between the UV illumination region and the oxygen exposure region indicates the significant effect of oxygen on the UV-treated PEDOT:PSS, explaining the importance of high vacuum for preserving the low work function of PEDOT:PSS.

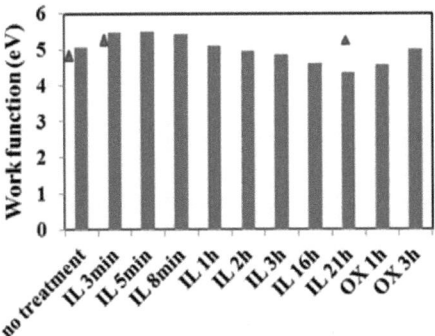

Figure 3.5 Comparison of work function of PEDOT:PSS after UV irradiation/oxygen exposure measured by field emission in vacuum. Triangles represent the work function value measured by photoelectron yield spectroscopy in air. (Adopted from Ref. 10.)

The absolute value of the work function measured by field emission may be not as accurate as those results measured via UPS, KB, and PYS. Nonetheless, the changing tendency of the work function of PEDOT:PSS under treatment may be verified by a different method. We measure the work function of PEDOT:PSS film by PYS. Similar enhancement after 3 min of UV treatment in vacuum (from 4.96 eV for as-prepared film to 5.08 eV for the 3 min sample) is obtained, confirming the reliability of the field emission technique. Different work functions are measured by field emission and PYS from PEDOT:PSS film after a long period of illumination (more than 10 h). Detailed analysis of the work function change under different atmosphere (in vacuum and in air) will be given in the next chapter.

Further evidence verifying the accuracy of the field emission method is provided in the estimation of the value of β. The radius of curvature of the tip was approximately 90 nm, as observed from Figure 3.1. We calculate 2.22×10^4 cm^{-1} as the value of β based on the shape of this tip ($k = 5$). This value is close to the result of 2.55×10^4 cm^{-1} calculated using Eq. (2) ($\varphi = 5.06$ eV). The concurrence of these values therefore further verifies the reliability of our measurements.

Work function has a critical influence on the contact performance in organic devices. *In situ* field emission measurement gives an alternative method to determine the work function and to investigate the contact behavior in future. Some contents in this chapter have been published in Ref. 10.

References

[1] Hong Ma, Hin-Lap Yip, Fei Huang, and Alex K.-Y. Jen, Interface Engineering for Organic Electronics, Adv. Funct. Mater. 2010, 20, 1371–1388

[2] J. Hwang, F. Amy, and A. Kahn, Spectroscopic study on sputtered PEDOT·PSS: Role of surface PSS layer, Org. Electron. 7, 387–396 (2006).

[3] T.-W. Lee and Y. Chung, Control of the surface composition of a conducting-polymer complex film to tune the work function Adv. Funct. Mater. 18, 2246–2252 (2008)

[4] S.-I. Na, G. Wang, S.-S. Kim, T.-W. Kim, S.-H. Oh, B.-K. Yu, T. Lee, and Dong-Yu Kim, J. Mater. Chem. 19, 9045–9053 (2009).

[5] A. M. Nardes, M. Kemerink, M. M. de Kok, E. Vinken, K. Maturova, and R. A. J. Janssen, Conductivity, work function, and environmental stability of PEDOT:PSS thin films treated with sorbitol, Org. Electron. 9, 727–734 (2008).

[6] T. Nagata, S. Oh, T. Chikyow, and Y. Wakayama, Effect of UV–ozone treatment on electrical properties of PEDOT:PSS film, Org. Electron. 12, 279–284 (2011).

[7] N. Koch, A. Vollmer, and A. Elschner, Influence of water on the work function of conducting poly(3,4-ethylenedioxythiophene)/poly(styrenesulfonate), Appl. Phys. Lett. 90, 043512 (2007).

[8] A. Modinos, *Field, Thermionic, and Secondary Electron Emission Spectroscopy* (Plenum Press, New York, 1984).

[9] Robert Gomer, Field emission and field ionization, New York: American Institute of Physics, 1993.

[10] Ying Jie Xing, Min Fang Qian, Jing Fang Qin, and Geng Min Zhang, Field emission study of change in work function of poly(3,4-ethylenedioxythiophene): poly(styrenesulfonate) film, Journal of Vacuum Science & Technology B 32, 02B101 (2014)

Chapter 4 Increased work function in PEDOT:PSS film under ultraviolet irradiation

4.1 Effect of UV treatment on work function of PEDOT:PSS film

As a necessary component in many organic devices, the stability of PEDOT:PSS film has been tested under various conditions. The effect of UV illumination on the performance of PEDOT:PSS film has been studied in air and a severe detriment on conductivity is observed under irradiation of short wavelength UV (wavelength less than 320 nm) [1]. It is therefore proposed that UV irradiation should be avoided for PEDOT:PSS film. A slight positive effect is reported when 365 nm UV is used as the light source, by which a small enhancement of conductive and work function is observed in PEDOT:PSS film [2].

UV ozone exposure is a common treatment to tune the surface property of substrates in organic electronics. Effect of UV ozone exposure on PEDOT:PSS film has been studied detailedly in Ref. 3. A large increase of 0.2 - 0.4 eV is measured after several minute of UV ozone treatment. The improvement of work function seems to obey a simple time-dependent law in this case. This phenomenon is reasonable because the work function enhances with the amount of oxidized bonds at the surface, which are produced monotonously with the treatment time of UV ozone. However, a shortcoming for this treatment is the enlargement of the film resistance after UV ozone exposure [3].

A claimed success in modifying PEDOT:PSS film is reported when PEDOT:PSS film after UV ozone exposure improves the performance of OLEDs effectively [4,5]. An enlargement of work function of PEDOT:PSS buffer layer after UV ozone treatment and a reduction of injection barrier are proposed to favor a better hole injection in OLEDs. Schematic energy diagram of UV ozone treated PEDOT:PSS/α-NPD is shown in Figure 4.1(b) according to Ref. 5. The energy diagram of untreated PEDOT:PSS/α-NPD is shown in Figure 4.1(a) for comparison. However, a doubt about the actual effect of this treatment is under debate. Using N,N′-diphenyl-N,N′-bis-(1-naphthyl)-1-1′-biphenyl-4,4′-diamine (α-NPD) as the hole transport material in a hole-only device, IV measurement shows that the hole injection from UV ozone treated PEDOT:PSS to α-NPD is depressed [6]. These conflicting results indicate that UV ozone may not be a proper treatment to reduce the barrier height in a practical fabrication process.

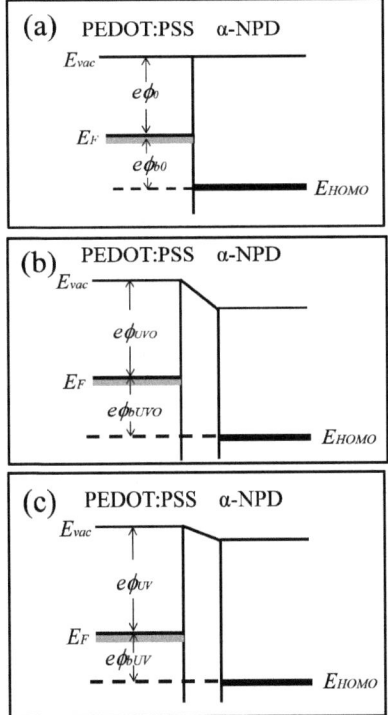

Figure 4.1 Schematic energy diagram of α-NPD and (a) untreated PEDOT:PSS, (b) UV ozone treated PEDOT:PSS, and (c) UV treated PEDOT:PSS. Vacuum energy alignment is assumed for untreated PEDOT:PSS/α-NPD interface. Both $e\phi_{UVO}$ and $e\phi_{UV}$ are larger than $e\phi_0$, $e\phi_{bUVO}$ is larger than $e\phi_{b0}$, $e\phi_{bUV}$ is less than $e\phi_{b0}$. (Adopted from Ref. 16.)

According to the results in above literatures [1-6], short wavelength UV irradiation and UV ozone treatment may increase the work function of PEDOT:PSS film by forming a metastable dipole layer at the surface, but decreases its conductance in the bulk. As discussed in Chapter 2, high work function and good conductivity are two most important aspects for PEDOT:PSS film as the anode buffer layer in organic optoelectronic devices. There still remains a challenge to improve the properties of PEDOT:PSS film by optimal UV treatment, which can increase the work function and conductivity in PEDOT:PSS film stably. We discuss our results on enhancing the work function of PEDOT:PSS film by 254 nm UV illumination in this chapter. The finding of conductivity improvement will be introduced in the next chapter.

4.2 Work function change under ambient atmosphere after UV irradiation

We use a commercial technique, photoelectron yield spectroscopy (PYS, Riken Keiki AC-2) [3,7], to evaluate the work function change in PEDOT:PSS film under ambient atmosphere after UV irradiation. PEDOT:PSS films are deposited on UV ozone-cleaned ITO glasses by spin coating and then bake at 160 °C in air for 20 min. Then the samples are fixed on a copper sample holder in the chamber of field emission microscope for UV irradiation (wave length 254 nm). The base pressure in this experiment is ~ 10^{-4} Pa. After different time of UV irradiation, the sample is taken from the vacuum chamber for PYS measurement in air. This method separates the process of UV illumination and oxygen exposure, providing a different way of UV treatment.

In photoelectron yield spectroscopy technique, a linear dependence between square root of photoelectron yield and the photon energy above the threshold is simulated to calculate the work function in metal, whereas for destination of highest occupied molecular orbital (HOMO) in semiconductor, the relation between a cubic root of photoelectron yield and the photon energy above the threshold is used instead. Generally, PEDOT:PSS is regarded as a metal in organic devices and work function measurement [3]. It is known that the work function of PEDOT:PSS is sensitive to the preparing condition. We find that there is a variation of 0.1−0.2 eV in different as-prepared PEDOT:PSS samples. This value is comparable to measured work function of PEDOT:PSS in literatures (4.9−5.2 eV) [8-10]. Two sets of samples are measured in order to determine the effect of illumination time on work function: 0, 3, 5 minute (short time) and 0, 10, 20 hour (long time). The data of PYS are shown in Figure 4.2. The values are 4.96, 5.08, 5.19 eV for 0, 3, 5 minute samples (Figure 4.2(a)), and 5.17, 5.48, 5.41 eV for 0, 10, 20 hour samples (Figure 4.2(b)), respectively. A universal increase of work function is observed from all samples in our experiments, which can be basically explained by an oxidation effect at the surface during ambient PYS measurement. The environment for PEDOT:PSS film changes from vacuum (pure and weak decomposition) to air (pure and weak oxidation) in PYS measurement. Decomposition effect of 254 nm UV light on chemical bonds has been studied for a long time [11]. The dangling state due to the cleavage of α-hydrogen bond in PSS may keep a long time in vacuum [12], and then, these broken bonds are sufficiently reactive for oxidation under ambient atmosphere, resulting in a more negatively charged surface and a positive space-charge layer beneath the surface [3]. The highest work function detected after 10 hour of UV irradiation indicates that the density of broken bonds reaches saturation at the surface.

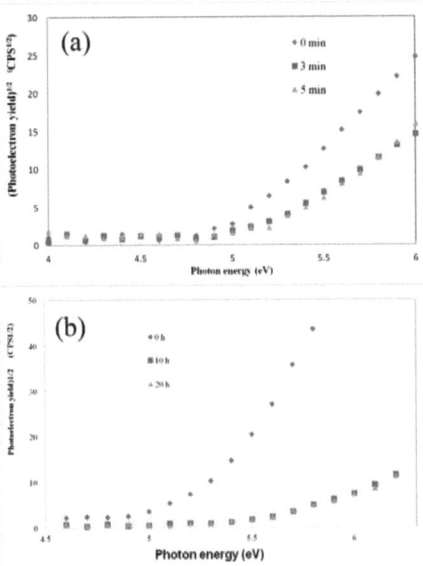

Figure 4.2 Photoelectron yield spectra of PEDOT:PSS films: (a) untreated (diamond), after 3 minute (square), and after 5 minute (triangle) of irradiation; (b) untreated (diamond), after 10 hour (square), and after 20 hour (triangle) of irradiation. (Adopted from Ref. 16.)

Although above oxidation effect is easy to understand, the time dependence of work function enhancement is not usual in our experiment. The power of UV light used here is much smaller than that used in UV ozone. However, only 5 minute of UV irradiation produced an obvious increase of 0.23 eV, which is close to the result of UV ozone treatment [3]. This fact gives a strong hint that the mechanism for work function improvement in our experiment cannot be described by a simple time dependent model.

27

PEDOT structure change can be excluded because no position shift and shape change are observed from Raman spectra in our experiments [2].

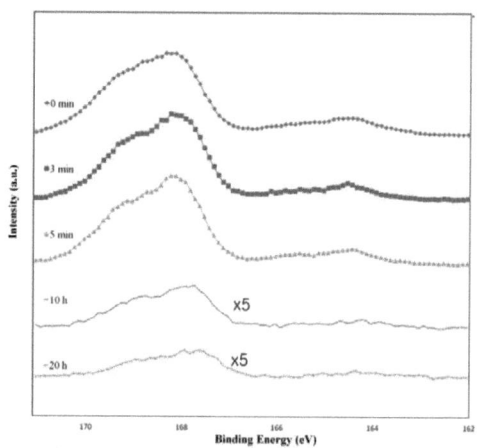

Figure 4.3 S 2p core spectra of PEDOT:PSS films: untreated, after 3 minute, 5 minute, 10 hour and 20 hour of irradiation. The intensity values of 10 hour and 20 hour samples are magnified with a factor of 5 in order to show the peak. (Adopted from Ref. 16.)

In general, the work function of as-prepared PEDOT:PSS film is highly related with PSS content at the surface. In order to verify the surface composition before and after UV treatment, XPS (Kratos Axis Ultra) analysis is used to determine PSS content. The peaks around 168 and 164 nm in XPS spectrum correspond to the sulfur atoms in PSS and PEDOT (S 2p peaks), respectively. The ratio of PSS peak area to PEDOT peak area in XPS plot can be used to evaluate the content at the film surface within several nanometers [8]. A series of PEDOT:PSS film after 3 minute, 5 minute, 10 hour, and 20 hour of UV treatment are analyzed by XPS and the S 2p spectra are shown in Figure 4.3. The ratios of S peak area (PSS to PEDOT) are calculated with the data in Figure 4.3. The values are 5.05, 6.58, 6.23 and 5.28 for no treatment, 3 minute, 5 minute, and 10 hour of UV treatment samples, respectively. These values reveal two phenomena in present experiments: first, the ratio of PSS/PEDOT increases significantly at the surface after a short time of UV irradiation; second, this ratio decreases as the illumination time prolonged. For the first phenomenon, we think that the detrimental effect on PEDOT by

UV treatment induces such a ratio enlargement. We know that the detrimental effect of 254 nm UV on PSS is stronger that of PEDOT. Nonetheless, both PSS and PEDOT can be decomposed simultaneously if the light intensity is high enough, as in the case of UV ozone treatment. It is well known that PEDOT content at the outmost surface (thin exposing layer in the top PSS-rich layer) is quite low because PEDOT's larger surface potential than that of PSS [13]. Therefore, if the outmost surface (where the light intensity is strongest) is decomposed indiscriminately by UV irradiation, even a small decomposition of PEDOT (reduction of absolute amount) may increase the PSS/PEDOT ratio (enlargement of relative ratio) evidently. In other words, at the very start of illumination, decomposition occurs without selectivity at the outmost surface. This assumption is coincident with the XPS result (Figure 4.3), in which the intensity of both PEDOT and PSS peak decrease after a short time UV treatment. The explanation for the second phenomenon is simple: longer UV irradiation breaks more bonds in PSS during the following period.

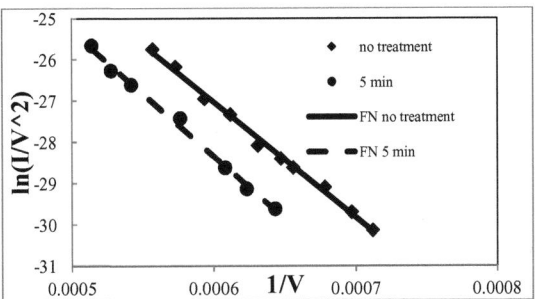

Figure 4.4 FN fitting lines of PEDOT:PSS without treatment and after 5 minute of UV illumination. (Adopted from Ref. 16.)

A remaining question for UV treatment is a small decrease of work function after 20 hour of irradiation (from 5.48 eV for 10 hour sample to 5.41 eV for 20 hour sample). This reduction of work function may come from the accumulating effect of the cleavage of sulfonic acid group by UV irradiation. It is reported that the work function of PSS wrapped carbon nanotubes drops 0.3 eV after strong UV illumination in vacuum because of the cleavage of the sulfonic acid groups [12]. In XPS spectrum of a sample after 20 hour of UV treatment (shown in Figure 4.3), the intensity of S 2p peak for PSS is rather weak, revealing that most sulfonic acid groups have been removed from PSS at the film surface. We think that this effect is much weaker than the influence of oxidation. Only in long-time irradiating samples, reduction of work function due to cleavage of sulfonic acid groups may be observed.

From above discussion, we may draw a conclusion that an improvement of work function will happen after a short period of UV irradiation even if without oxidation. This conclusion is proved by results of field emission measurements in Chapter 3, which has no influence of oxygen. F-N plots before and after UV irradiation are shown in Figure 4.4. Using 4.96 eV (measured by PYS) as the work function of untreated PEDOT:PSS film, the work function of PEDOT:PSS film is calculated as 5.37 eV after 5 minute of UV irradiation. Although a small deviation appears between PYS (5.19 eV) and field emission (5.37 eV), the increasing tendency of work function is the same in these two different experiments. The reason for this deviation is not clear. One possible reason may be the variation of work function in different PEDOT:PSS samples. Taking all above discussion, we believe that two mechanisms work for the improvement of work function after UV illumination: increasing PSS/PEDOT ratio and the oxidized surface. The first mechanism shows apparent effect after short time of UV illumination in vacuum and the second one plays a key role in air.

4.3 Performance of UV treated PEDOT:PSS film in organic devices

We fabricate hole-only devices to testify the effect of UV treatment. The device structure is schematically drawn in the inset of Figure 4.5. PEDOT:PSS layer treated after 0, 10, and 22 hour of UV irradiation are used in these devices. In order to preclude a possible build-in potential generated by organic-electrode contact [14], we deposit a thick layer (500 nm) of α-NPD on PEDOT:PSS layer to construct a hole-only device. Figure 4.5 shows the measured IV curves of three devices. It is clear that there are two regions in the plots: space charge limited region (>1 V) and injection limited region (< 1V). The well overlap of three curves in space charge limited region reveals that PEDOT:PSS layer treated by different conditions does not influence the charge transport under larger bias voltage in these devices, which is different from the results of UV Ozone treatment [4,6].

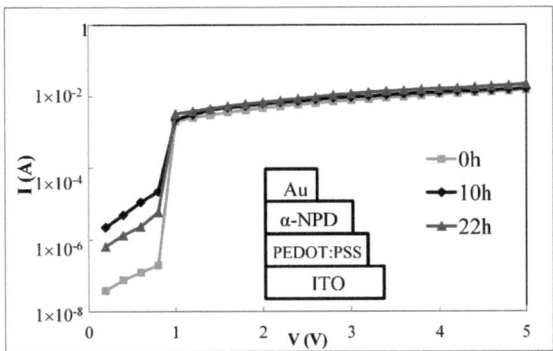

Figure 4.5 IV curves of hole-only devices. Inset: device structure. (Adopted from Ref. 16.)

A distinct impression from IV curves in injection limited region is that a larger work function of PEDOT:PSS corresponds to a larger current in device, indicating the injection barrier is reduced by UV illuminated PEDOT:PSS. However, more attention should be paid to the surface state of PEDOT:PSS, which enhances the height of injection barrier in device employing UV ozone treated PEDOT:PSS [4,6]. Hole injection from PEDOT:PSS to α-NPD layer may be modeled with a thermally activated mechanism and the detailed analysis of IV characteristic is described according to this model in Ref. 15. The dependence of current density on the barrier height in the injection limited region is described as

$$J \propto \exp\left(-\frac{e\phi_b}{k_B T}\right)$$

where $e\phi_b$ is the barrier height, e is the electron charge, k_B is the Boltzmann constant, and T is the temperature [15]. We estimate the effect of UV treatment on device performance by this method. A reduction of barrier height ($\Delta e\phi_b$) is calculated with the data in Figure 4.5 and the values are 0.11 and 0.07 eV for 10 and 22 hour sample, respectively. These values are smaller than the improvement of work function (ΔE_F), 0.31 and 0.24 eV for 10 and 22 hour sample, respectively. This difference reveals that, for UV treated PEDOT:PSS as a hole injection layer, the height of injection barrier is not only controlled by the work function difference, but also by the surface state of PEDOT:PSS. A dipole should exist between UV treated PEDOT:PSS and α-NPD ($\Delta E_F - \Delta e\phi_b$, 0.20 and 0.17 eV for 10 and 22 hours sample, respectively), and this dipole counteracts part effect of the improvement of work function of UV irradiated PEDOT:PSS film. Schematic energy diagram of UV treated PEDOT:PSS/α-NPD is shown in Figure 4.1(c). All measured data and calculated results are summarized in Table 4.1.

Table 4.1 Work functions measured by PYS and field emission, calculated change of barriers height and dipole, in PEDOT:PSS film under short and long time illumination.

	Illumination time	PYS result (eV)	ΔE_F (PYS) (eV)	Field emission result (eV)	ΔE_F (Field emission) (eV)
Short time illumination	0 min	4.96	–	4.96	–
	3 min	5.08	0.12	–	–
	5 min	5.19	0.23	5.37	0.41
Long time illumination	Illumination time	PYS result (eV)	ΔE_F (PYS) (eV)	$\Delta e\phi_b$ (eV)	$\Delta E_F - \Delta e\phi_b$ (eV)
	0 h	5.17	–	–	–
	10 h	5.48	0.31	0.11	0.20
	20 h	5.41	0.24	0.07	0.17

A different behavior between UV treated and UV ozone treated PEDOT:PSS film is observed with same device structure. By separating UV decomposition and oxidation, a lower injection barrier is achieved in a hole-only device, revealing the advantage of UV irradiation over UV ozone exposure. Other than oxidation effect, enhancing PSS/PEDOT ratio is determined as the other independent mechanism for work function improvement under UV irradiation. We suggest that this content variation causes the reduction of barrier height, which is not influenced by the interfacial dipole. Some contents in this chapter have been published in Ref. 16. We believe that our result provides a better way to control the interface between PEDOT:PSS and organic semiconductor.

References

[1] Andreas Elschner, Stephan Kirchmeyer, Wilfried Lövenich, Udo Merker, Knud Reuter, PEDOT: principles and applications of an intrinsically conductive polymer, Boca Raton: CRC Press, 2011

[2] Lin Y, Yang F, Huang C, et al. Increasing the work function of poly (3,4-ethylenedioxythiophene) doped with poly(4-styrenesulfonate) by ultraviolet irradiation. Appl Phys Lett, 2007, 91(9): 092127

[3] Nagata T, Oh S, Chikyow T, et al. Effect of UV–ozone treatment on electrical properties of PEDOT:PSS film. Org Electron, 2011, 12(2): 279–284

[4] Benor A, Takizawa S, Perez-Bolivar C and Anzenbacher P 2010 Organic Electronics 11 938

[5] Benor A, Takizawa S, Chen P, Perez-Bolivar C and Anzenbacher P 2009 APPLIED PHYSICS LETTERS 94 193301

[6] Helander M G, Wang Z B, Greiner M T, Liu Z W, Lian K and Lu Z H 2009 APPLIED PHYSICS LETTERS 95 173302

[7] Lee T W and Chung Y 2008 Adv. Funct. Mater. 18 2246

[8] Na S I, Wang G, Kim S S, Kim T W, Oh S H, Yu B K, Lee T and Kim D Y 2009 J. Mater. Chem. 19 9045

[9] Nardes A M, Kemerink M, de Kok M M, Vinken E, Maturova K, Janssen R A J 2008 Organic Electronics 9 727

[10] Lee H K, Kim J K and Park O O 2009 Organic Electronics 10 1641

[11] Rice R G 1982 Handbook of ozone technology and applications(Ann Arbor: Science)

[12] Hong K, Kim S H, Yang C, Yun W M, Nam Si, Jaeyoung Jang, Park C and Park C E 2011 ACS Appl. Mater. Interfaces 3 74

[13] Hwang J, Amy F and Kahn A 2006 Organic Electronics 7 387

[14] Helander M G, Wang Z B, Greiner M T, Qiu J and Lu Z H 2009 REVIEW OF SCIENTIFIC INSTRUMENTS 80 033901

[15] Wang Z B, Helander M G, Greiner M T, Qiu J, and Lu Z H 2010 JOURNAL OF APPLIED PHYSICS 107 034506

[16] Xing Yingjie, Qian Minfang, Guo Dengzhu, and Zhang Gengmin, Increased work function in PEDOT:PSS film under ultraviolet irradiation, Chin. Phys. B Vol. 23, No. 3 (2014) 038504

Chapter 5 UV irradiation induced conductivity improvement in PEDOT:PSS film

5.1 Conductivity of PEDOT:PSS film

Thermally stable PEDOT:PSS film formed by spin coating or inkjet printing has a structure of horizontal layers of PEDOT-rich clusters separated by insulating PSS lamella [1], which is shown in Figure 5.1(a). The PSS lamella hinders charge transport severely and causes an anisotropic conductance inside the film. 10^{-3} S/cm for parallel direction (parallel to the film surface) and 10^{-6} S/cm for perpendicular direction (perpendicular to the film surface), are measured in as-prepared PEDOT:PSS film [1]. The conductivity of PEDOT:PSS film shows a direct influence on the performance of organic optoelectronic device, for example, efficiency and life time of polymer solar cell are enhanced by employing PEDOT:PSS film with high conductivity [2–4].

A common technique to enhance the conductivity of PEDOT:PSS film is by addition of small amount of polar solvent or high boiling point alcohol into the commercial PEDOT:PSS solution before spin coating, such as dimethylsulfoxide [2], glycerol [5], and sorbitol [6]. The reason for conductivity improvement is the formation of a uniformly conductive network in PEDOT:PSS film, which is evidenced by the aggregation of PEDOT-rich particles observed by atomic force microscopy (AFM) [2,6]. Post-treatment is another method to increase the conductivity after PEDOT:PSS film formation. Kim et al. report a large conductivity increase by immersing as-prepared PEDOT:PSS film in an ethylene glycol bath [7]. A very high conductivity (3065 S/cm) is obtained in H_2SO_4 post-treated and water-rinsed PEDOT:PSS film (with Clevios PH1000 for electrode application) [8]. Effectively replacement of PSS$^-$ ions with HSO_4^- causes phase separation between PSS and PEDOT chains, resulting in an ITO– comparable conductivity. Much less PSS means much less transport barrier around PEDOT-rich clusters in post-treated PEDOT:PSS film. Up to date, most progresses in improving the conductivity of PEDOT:PSS film adopt above two approaches, resulting in more uniform PSS redistribution in both parallel and perpendicular direction or PSS removal by solution addition or post-treatment, respectively.

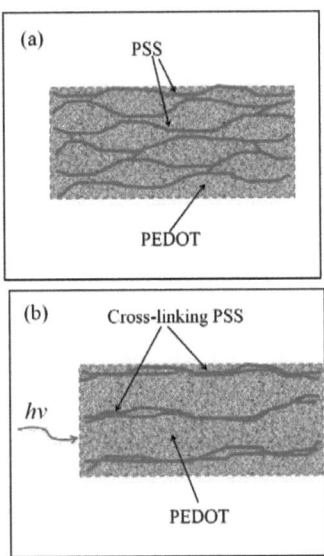

Figure 5.1 Cross-sectional view of the schematic morphological model for (a) as-prepared PEDOT:PSS film; (b) UV illuminated PEDOT:PSS film in vacuum, PEDOT-rich particles (gray) connects each other in parallel direction without PSS obstacles due to cross-linking PSS chains (blue curves) after long time of UV irradiation. (Adopted from Ref. 15.)

The effect of UV treatment on conductivity of PEDOT:PSS film has been studied and the result is not satisfactory. It is known that UV illumination may destroy organic compounds. Specifically, no decomposition occurs under 365 nm UV irradiation, whereas 254 nm UV light can break many kinds of chemical bond, such as C-C, C-H, and C-OH [9]. For PEDOT:PSS film, PSS is more detrimental under 254 nm UV irradiation than PEDOT because π-conjugated part in PEDOT chain has a larger bonding energy than single bond in PSS chain [10]. It is therefore expected that selective inhibition of PSS barrier in PEDOT:PSS film may be realized by proper UV irradiation. However, a reduced conductivity is measured in PEDOT:PSS film illuminated by short wavelength UV (wavelength less than 320 nm) [11], and only a slight conductivity improvement in PEDOT:PSS film is observed under 365 nm UV illumination [12]. UV ozone treatment also causes a significant increase of resistance in PEDOT:PSS film [10].

These experiments indicate that short wavelength UV treatment under ambient atmosphere shows no beneficial effect on transport property of PEDOT:PSS film.

5.2 Conductivity enhancement in PEDOT:PSS film under UV irradiation

We treat PEDOT:PSS film with 254 nm UV in a different way, by which PEDOT:PSS film is illuminated in vacuum. No oxidation occurs in PEDOT:PSS film during UV irradiation in this case. ITO glass and Si plate covered with SiO_2 layer (thickness 300 nm) are used as the substrate for perpendicular and parallel conductivity measurement, respectively. Commercial PEDOT:PSS solution is used without any solvent addition in our experiment. After spin coating, PEDOT:PSS film is baked at 160 °C for 20 min in air. Surface profiler measurement shows that PEDOT:PSS film has a thickness of about 45 nm. Then the sample is fixed on a copper sample holder in a stainless steel vacuum chamber. The sample holder is connected directly with chamber to maintain the sample temperature at room temperature. The base pressure of the chamber is ~10^{-4} Pa. The effect of UV irradiation on conductance of PEDOT: PSS film is studied within the vacuum chamber.

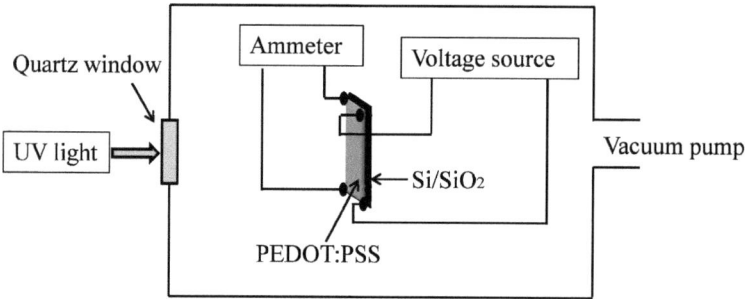

Figure 5.2 Schematic configuration of the setup for measuring parallel conductivity of PEDOT:PSS in vacuum.

The parallel conductivity is measured by van der Pauw four-point probe technique. Small In pad is used as the contact at four corners of a PEDOT:PSS/SiO_2/Si sample (1×1 cm^2). A schematic configuration of the measuring setup is shown in Figure 5.2. The calculation method can be found in Ref. 8. Square gold electrodes (2×2 mm^2) are deposited on PEDOT:PSS/ITO by thermal evaporation for perpendicular conductivity measurement. A pair of parallel Ti/Au electrodes (length 5 mm) with the distance of 5 mm are deposited on SiO_2/Si for addition planar IV measurement before PEDOT:PSS deposition. Electrode thicknesses of 50 and 75 nm are chose for perpendicular and planar conductance measurement, respectively.

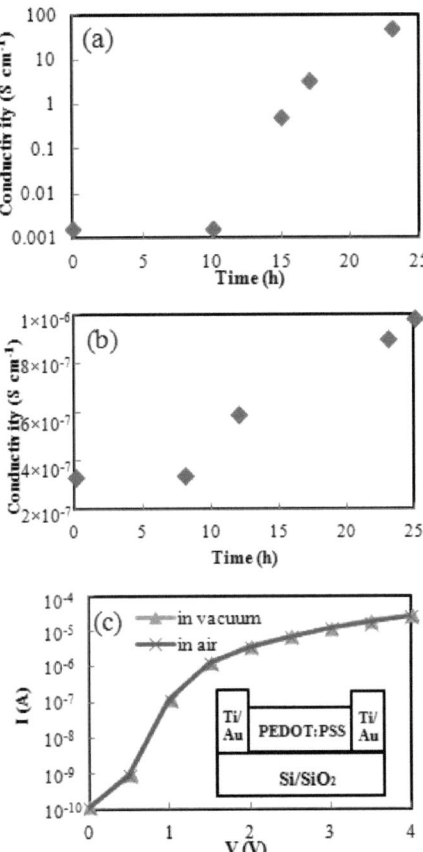

Figure 5.3 Treatment time dependence of (a) parallel and; (b) perpendicular conductivity of PEDOT:PSS film; (c) IV curves of a planar device after 23 h of UV illumination measured in vacuum and in air. Inset: device structure for planar conductance measurement. (Adopted from Ref. 15.)

The untreated sample is measured before (in air) and after evacuation (in vacuum) with same electrical connections firstly. Then a UV light is used to irradiate the sample in vacuum. After a period of irradiation, the quartz window is covered by a light barrier and IV measurement is conducted in the dark. This procedure is repeated by more than 10 times. No difference due to environment change is detected in untreated sample and the measured conductivity (1.73×10^{-3} S/cm) is in accordance with the value in Ref. 1. Usually, a slight increase of conductivity is found after about 10 h of irradiation. More conductivity enhancement can be achieved by longer irradiation until the improvement tends to become saturated after about 21 h of irradiation. The treatment time dependence of the parallel conductivity is summarized in Figure 5.3(a). After 23 h of irradiation, the conductivity becomes 50.47 S/cm, which means a dramatically improvement by four orders. This value is higher than the result (10 S/cm) obtained by sorbitol addition, which is the best conductivity of PEDOT:PSS film in literatures (with Baytron P VP AI 4083) [6]. The stability of improved conductivity is tested by measuring the UV treated sample in air. IV measurement reveals no conductivity reduction occurs during 10 h of air exposure. Additional device in planar structure is fabricated and treated in the same way to confirm this behavior of PEDOT:PSS film. The device structure is schematically drawn in the inset of Figure 5.3(c). Significant conductance improvement is observed after 23 h of UV irradiation. The stability of UV illuminated planar device is checked by IV measurement in vacuum and then in air. No difference due to environment variation is detected under small bias (shown in Figure 5.3 (c)). Only a slight conductance reduction occurs under bias voltage of 4 V after several h of ambient exposure (in air 2.65×10^{-5} A, in vacuum 2.70×10^{-5} A). We note that by naked eye, no color or transparency change in treated PEDOT:PSS film is observed.

We also examine the perpendicular conductivity of PEDOT:PSS film before and after UV irradiation. The device structure is ITO/PEDOT:PSS/Au for vertical IV measurement. The treatment time dependence of the perpendicular conductivity is shown in Figure 5.3(b). The perpendicular conductivity of untreated PEDOT:PSS film is 3.35×10^{-7} S/cm, which is a little smaller than the reported value in Ref. 1. This difference should come from the larger series resistance of ITO electrode in our experiment. The perpendicular conductivity becomes 9.81×10^{-7} S/cm after 25 h of irradiation, indicating that anisotropic conductivity still remains in UV treated

PEDOT:PSS film. All data of Figure 5.1(a) are listed in Table 5.1. All data of Figure 5.1(b) are listed in Table 5.2.

Table 5.1 PEDOT:PSS parallel conductivity after UV illumination.

Illumination time (hour)	Conductivity (S/cm)
0	1.73E-03
10	1.74E-03
15	5.36E-01
17	3.70E+00
23	5.05E+01

Table 5.2 PEDOT:PSS perpendicular conductivity after UV illumination.

Illumination time (hour)	Conductivity (S/cm)
0	3.35E-07
8	3.39E-07
12	5.91E-07
23	8.98E-07
25	9.81E-07

Above results show clearly that 254 nm UV illumination in vacuum has significant effect on the conductance of PEDOT:PSS film. Some additional experiments are performed to verify this conclusion. We cannot detect any resistance change even after 30 h of irradiation when 365 nm UV light is used instead of 254 nm UV. On the other hand, only a small change in conductivity is found after 254 nm UV illumination when PEDOT:PSS film is placed in air.

5.3 Mechanism for conductivity improvement in PEDOT:PSS film

The mechanism for conductivity improvement induced by 254 nm UV irradiation is a little confusing. Up to date, most efficient technique to enhance the conductivity of PEDOT:PSS film is to redistribute or remove PSS by solution-involved method. However, there is no driving force for PSS redistribution inside a solid film normally.

Annealing effect due to long time illumination can be excluded from our experiment because the sample temperature does not change. Ouyang et al. propose a mechanism for conductivity enhancement due to the structure change of PEDOT chains [14], in which formation of quinoid conformation in PEDOT chains is supposed and a shift of Raman peak around 1440 cm^{-1} is used as the pointer for the conformation of quinoid structure in PEDOT chains. We investigate the Raman spectra of PEDOT:PSS films before and after 254 nm UV treatment. No position shift and shape change are observed in these samples (shown in Figure 5.4). This result reveals that no structure change of PEDOT chain happens after UV irradiation in our experiment.

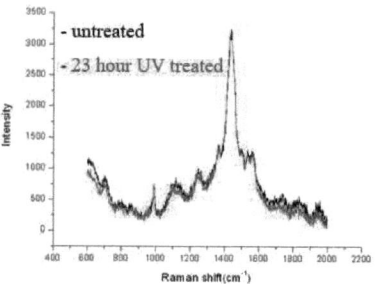

Figure 5.4 Raman spectra of PEDOT:PSS film before and after 23 h of UV illumination. (Adopted from Ref. 15.)

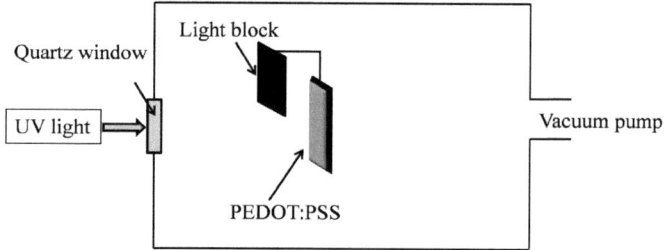

Figure 5.5 Schematic configuration of the setup to distinguish the effect of UV illumination and vacuum storage on PEDOT:PSS.

More experiments are carried out inside the vacuum chamber to clarify the mechanism for conductivity increase. A small light barrier is mounted in front of one part of PEDOT:PSS film (not contact the film surface) to block the UV light. In this way, we can distinguish the effect of UV irradiation from vacuum storage. Schematic configuration of this setup is shown in Figure 5.5.

Two samples are prepared in the vacuum chamber: 5 min (short time illumination) and 10 h (long time illumination) of UV irradiation. Surface profiler examination shows that no significant thickness difference exists between exposed and shaded region. The surface morphology of PEDOT:PSS is scanned by AFM. The images of 5 min sample are demonstrated in Figure 5.6(a) and no obvious difference is found in these two figures. The values of root mean square (RMS) roughness are 0.89 nm (shaded region) and 0.92 nm (illuminated region), respectively. This phenomenon means that the film surface is a little roughened after a short period of UV irradiation. Drastic morphology change happens after 10 h of UV irradiation. Comparing to the shaded surface (RMS 0.94 nm, left image of Figure 5.6(b)), small particles and elongated structures are generated on the surface of irradiated part (right image of Figure 5.6(b)). From similar morphology of shaded part in 5min and 10 h sample, it is obvious that no morphology change is produced by vacuum storage. This result confirms that the increased conductivity after UV irradiation cannot come from PSS redistribution/removal, indicating the existence of other mechanism for conductivity improvement in our experiment.

A noticeable characteristic of our experiment is that there seems a time threshold for detectable effect of UV irradiation. Morphology change due to 10 h of UV illumination (shown in Figure 5.6(b)) is in accordance with the occurrence for conductivity rising, indicting a relationship between film structure and charge transport. Decomposition effect of 254 nm UV light on polymer has been well-studied in literature [9]. A special condition in our experiment is vacuum environment during UV treatment, in which a divergent effect due to detrimental difference between PEDOT and PSS can be accumulated and observed. Hong et al. have studied the effect of 254 nm UV illumination on pure PSS film in vacuum [14]. They report that two kinds of bond in PSS chain will be broken under UV illumination. The cleavage of a α-hydrogen bond generates polystyrene radicals, and the second cleavage happens between sulfonic acid group and phenyl group. Different from oxidation caused by 254 nm UV irradiation in air, a benzene ring forms immediately after the cleavage of the second kind of bond (a fast reaction) and a slow cross-linking among polystyrene radicals occurs after 40 min of illumination (a continuous process) [14]. According to their conclusion, we explain the conductivity improvement in our experiment as following. During the first several hour of UV irradiation, broken α-hydrogen bonds keep their dangling state in vacuum and their amount goes up monotonously with the illumination time. When the intensity of these cleaved bonds is high enough (after about 10 h of UV illumination in our experiment), photo-cross-linking of PSS chain happens among UV-generated radicals. In the period of 20 h of UV irradiation, such a cross-linking occurs at the film surface firstly, and then takes place in the bulk of PEDOT:PSS film. This suggestion is reasonable since the existence of vertical phase separation and much more absorbance of PSS than PEDOT at this short wave length [7].

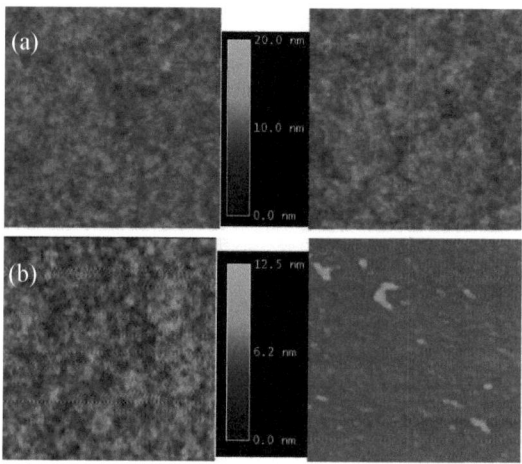

Figure 5.6 (a) AFM images of PEDOT:PSS film after 5 min of illumination; (b) AFM images of PEDOT:PSS film after 10 h of illumination, Left and right image in (a) and (b) shows shaded and irradiated region, respectively. (Adopted from Ref. 15.)

From above discussion, we propose that PSS cross-linking in PEDOT:PSS film alters film structure and lessens transport barrier between parallel neighboring PEDOT-rich particles effectively. Such a slight change of PSS chain conformation is shown in Figure 5.1 schematically. Some continuous PEDOT-rich routes are formed inside the PEDOT:PSS film after 10 h of UV irradiation (Figure 5.1(b)), and the structure model of as-prepared PEDOT:PSS film is shown in Figure 5.1(a) for reference. As the result of anisotropic PEDOT-rich routes, the increased conductivity after UV irradiation occurs mainly in parallel direction, which is in contrast to the isotropic improvement of conductivity achieved by solvent addition [2]. In order to prove PSS cross-linking after UV irradiation, we analyze UV treated PEDOT:PSS film by optical methods. However, no obvious peak is found in Fourier transform infrared (FT-IR) spectra due to thin thickness of PEDOT:PSS film. Then we measure the absorption spectrum of PEDOT:PSS film prepared on quartz substrate. UV absorption is always used as the gauge for detecting PSS content in PEDOT:PSS film because PSS has strong absorption in UV band [8,14]. As shown in Figure 5.7, small variation of UV absorption is observed in 10 h sample, and stronger UV absorption is found in 20 h sample. This result indicates effective cross-linking of PSS chains after 20 h of UV illumination [14]. The reason for saturation of conductivity increase after about 21 h of irradiation is still not clear. One possible explanation is that a small amount of PEDOT is also destroyed by UV irradiation, resulting in some less-conductive routes after long time irradiation.

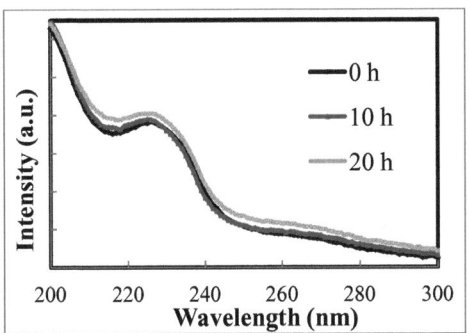

Figure 5.7 UV absorption spectra of PEDOT: PSS film before and after 10 and 20 h of UV illumination. (Adopted from Ref. 15.)

A shortcoming of our method is the long irradiation time to achieve the largest conductivity. This problem can be overcome by enlarging the power of UV light source, because more bonds will be broken under strong UV exposure as in Ref. 14. This approach may bring a worry about the stability of PEDOT:PSS film after stronger UV treatment, since some broken bonds in PEDOT:PSS may be oxidized in air due to insufficient PSS cross-linking among high intensity radicals. We propose that the stability of UV treated PEDOT:PSS in air can be evaluated by conductivity improvement in vacuum, because the occurrence of conductivity improvement indicates the start of elimination of α-hydrogen cleavage via cross-linking and most broken α-hydrogen bonds disappear when film conductivity reaches the peak value. This mechanism proves that a stable sample can be obtained by monitoring the conductivity change. Therefore, we always obtain air-stable sample in our experiment. In other words, higher conductivity PEDOT:PSS film shows, more stable the sample is. An additional benefit of our method is the possibility of tuning the conductivity of PEDOT:PSS film independent of film formation process. For example, our method may be compatible with roll-to-roll fabrication of flexible device. We believe that our result provides more information for treating PEDOT:PSS in the future.

In this work, we obtain up to four order of conductivity increase (from 10^{-3} S/cm in as-prepared film to 50 S/cm in UV treated film) by irradiating PEDOT:PSS film with 254 nm UV. This result is better than the highest value of parallel conductivity (10 S/cm) of PEDOT:PSS film in literature (with Baytron P VP AI 4083) [9]. Vacuum is found as the necessary environment for UV treatment. Less barrier due to cross-linking among UV damaged PSS chains is determined as the reason for improved charge transport. Some contents in this chapter have been published in Ref. 15. Our result provides a novel route to enhance the conductivity of PEDOT:PSS film other than solution method.

References

[1] Nardes A, Kemerink M, Janssen R, et al. Microscopic understanding of the anisotropic conductivity of PEDOT:PSS thin films. Adv Mater, 2007, 19(9): 1196–1200

[2] Na S, Wang G, Kim S, et al. Evolution of nanomorphology and anisotropic conductivity in solvent-modified PEDOT:PSS films for polymeric anodes of polymer solar cells. J Mater Chem, 2009, 19(47): 9045–9053

[3] Kim Y, Ballantyne A, Nelson J, et al. Effects of thickness and thermal annealing of the PEDOT:PSS layer on the performance of polymer solar cells. Org Electron, 2009, 10(1): 205–209

[4] Kim Y, Sachse C, Hermenau M, et al. Improved efficiency and lifetime in small molecule organic solar cells with optimized conductive polymer electrodes. Appl Phys Lett, 2011, 99(11): 113305

[5] Snaith H, Kenrick H, Chiesa M, et al. Morphological and electronic consequences of modifications to the polymer anode 'PEDOT:PSS'. Polymer, 2005, 46(8): 2573–2578

[6] Nardes A, Kemerink M, Kok M, et al. Conductivity, work function, and

environmental stability of PEDOT:PSS thin films treated with sorbitol. Org Electron, 2008, 9(5): 727–734

[7] Kim Y, Sachse C, Machala M, et al. Highly conductive PEDOT:PSS electrode with optimized solvent and Post-treatment for ITO-free organic solar cells. Adv Funct Mater, 2011, 21(6): 1076–1081

[8] Xia Y, Sun K, Ouyang J. Solution-processed metallic conducting polymer films as transparent electrode of optoelectronic devices. Adv Mater, 2012, 24(18): 2436–2440

[9] Rice R. Handbook of ozone technology and applications. Ann Arbor: Ann Arbor Science, 1982

[10] Nagata T, Oh S, Chikyow T, et al. Effect of UV–ozone treatment on electrical properties of PEDOT:PSS film. Org Electron, 2011, 12(2): 279–284

[11] Andreas Elschner, Stephan Kirchmeyer, Wilfried Lövenich, Udo Merker, Knud Reuter, PEDOT: principles and applications of an intrinsically conductive polymer, Boca Raton: CRC Press,

[12] Lin Y, Yang F, Huang C, et al. Increasing the work function of poly (3,4-ethylenedioxythiophene) doped with poly(4-styrenesulfonate) by ultraviolet irradiation. Appl Phys Lett, 2007, 91(9): 092127

[13] Ouyang J, Xu Q, Chu C, et al. On the mechanism of conductivity enhancement in poly(3,4-ethylenedioxythiophene):poly(styrene sulfonate) film through solvent treatment. Polymer, 2004, 45(25): 8443–8450

[14] Hong K, Kim S, Yang C, et al. Photopatternable poly(4-styrene sulfonic acid)-wrapped MWNT thin-film source/drain electrodes for use in organic field-effect transistors. ACS Appl Mater Interfaces, 2011, 3(1): 74–79

[15] XING YingJie, QIAN MinFang, WANG GuiWei, ZHANG GengMin,GUO DengZhu and WU JinLei, UV irradiation induced conductivity improvement in poly(3,4-ethylenedioxythiophene):poly(styrenesulfonate) film, Sci China Tech Sci, 2014, 57: 4448